Shipping, Maritime and Ports in India

By
Chakrapani Srinivasa

Shipping, Maritime and Ports in India
By
Chakrapani Srinivasa

Dedicated to my dear parents

Preface

Indian shipping, Ports and Maritime industry have seen many uplifts recently.

Steps to be taken for betterment in these fields are to be attended.

Building of new ports and shipyards, efforts in sustainability in Indian Ports, improvement in coastal and inland waterways, International Conferences on shipping and maritime expo details are exposed explicitly in this book.

Contents

Efforts in Sustainability of Indian Ports and Port Management

The Port Management needs a turn around.

"The CEOs in all Indian Ports should support safe and effective usage of technologies.

Steps should be taken to update basic navigational and engineering skills. They should analyze risks, minimize it and implement digitally. Total Quality Management and Risk Management should be their routine job. They should be prepared to welcome ships with unmanned engine rooms. Port tragedies should enable them to gain experience and prevent recurrence of it" says a Maritime expert.

VDR data is needed after accidents. AIS and UTTS for collision avoidance is a must. Knowledge on Navigation Audit is needed to maintain the standard of Port.

"New design of Life Boat hooks will reduce accidents. All data required for smooth operation of a Port are not available. Further, presentation of data is confusing. So, totally Port Management needs a new look.

ESID, EVC Charts and PSC inspection are considered important for better functioning of ports and vessels operated there.

Ship operators should be encouraged to conduct Risk Assessment on board operations" says a naval officer.

Container Gate Transaction

Ports should have highly talented youths. Hence the government should support bright young professionals in leading universities and sea farer training institutions. Certification of Academic Institutions will produce qualified personals for Indian Ports. Work sharing is to be done by digital revolution.

The introduction of ERP system is a major turnaround in Indian ports for its sustainability. Now it is in existence in Vizakapattinam, Mangalore Port, VOC, Tutucorin Port, and Mormugao. The Port Community System to obtain seamless connectivity between various points is a master piece in major ports of India.

To avoid backlog in cargo evacuation, the introduction of Container Gate Transaction is of great importance.

Acquire Ships Overseas

Abatement of service tax from 50% to 60% is done now. With this change as per a budget only 40% of the value of service for transportation of goods by vessels will be subjected to levy of service tax.

Merchandise trade in India is low upto 42%, which has to be enhanced without delay as other countries have achieved 75%. 34 projects are on the anvil in major ports with Public Private Partnership mode with a huge investment of Rs.1150 crores. Up gradation of software applications in ports is a vital step for sustainability. The new government has generously framed a policy to encourage shipping enterprises in India to acquire ships overseas and flag them in India as per their whims and fancies. The government will not interfere in their decision making system with constraints like religion, caste and language.

Tax and Subsidy

The exemption of Customs and Excise Duty on bunker fuels utilized in Indian flag vessels for transportation of EXIM and empty containers are heartily welcomed by Port authorities.

The user friendly Trade Portal developed by Federation of Indian Export organization is another step of fortune for port sector.

The Reserve Bank of India and Ministry of Finance have been directed by the new government to sanction a special dispensation for 5 years upto 2020, to treat Repeat Restructuring of Shipyards after failure of CDR, as equivalent to first restructuring.

The extension of ship building subsidy scheme of 2002-2007 from Oct 2009 till March 31st 2014 for liquidation of committed liabilities for ship building contracts obtained during 2002-2007 is being done by government of India.

Market Linked Tariff System

The Tariff reforms and modifications in official procedures have encouraged the investment of Rs.20000 crores towards 30 infrastructure projects recently. Now they have even increased to Rs.20.709 crores to reach 218 MTPA capacity.

Smaller projects were also planned to the tune of Rs.6765 crores to gain a capacity of 136.75 MTA additionally. A trend has been created to intensify competition in the field due to modifications in this new Market Linked Tariff system. This aided the erection of mega container terminal at Tuna – Tekra Kandla Port with an investment of Rs.6000 crores, along with other projects in other States.

Scenario should change

Though guidelines have been changed, the government has to probe into this issue deeply to attract global investors with sophistication in operation of ports.

Poor rail links is a vital bottleneck for port sustainability. For example in handling of coal, absence of rail links has made Coal India unable to supply 500 mt of additional coal, says a press report. This has to be viewed seriously and more budget allocation for roads and railways is needed. States need vital railway projects for linking the ports.

A healthy scenario is a must with better infrastructure.

————————————————————————

Ship Logistic / Supply Chain

The changes are required to develop a working model for India to become a sustainable successful Maritime State.

"We face difficulties in Logistics Sector, as we need 46 documents, whereas in China, every formality is cleared through 4 papers" said an industrialist.

Delay kills logistics

"India has to tone up in various aspects including ship Logistic / Supply Chain.

Maritime sector is not considered as seriously like road and rail sector. Industrial development and GDP growth fully depends upon good logistics and supply chain" says an economist.

As said by Chief of Naval Staff, R.K. Dhowan in INMARCO conference, most of the fish along the Indian shore die due to old age. There is no one to catch it as the fishermen fear that the efforts they put into catch them will not fetch them results. They cannot take them to the market in a proper fashion.

"Due to poor infrastructure, ill maintained roads and weak logistics system, many valuable goods like fish, grains, rice, cotton, sugarcane etc., remain a waste for Indian economy" said he.

As said by Shantanau Paul, General Manager of Shipping Corporation of India, the transport charge to carry cotton Gujarat to Tamilnadu is 4 times more than the charges to carry the same cotton from India to China. We face difficulties in Logistics Sector, as there is a demand for 46 documents, whereas in China, every formality is cleared through 4 papers. This drastic delay kills Logistics and Maritime industry.

The road sector and railway division have enough manpower and financial support, but Maritime Industry faces poor allocation of funds and lack of skilled man power. We have to dream of a good sustainable and successful maritime state.

Want of Man Power

Many do not prefer marine course as the family structure, beliefs and pattern of living of Indians is so sensitive that they do not allow their sons to be away from family and dwell in the sea.

The family bonds prevent the entry of Indian youths to Maritime sector. So, all the activities concerned with it are held up for want of manpower. The Supply Chain sector is to be upgraded to make India a fortunate Maritime State.

Experts from developed countries should join hands in our Logistics System to make it sophisticated and fast.

Uniform GST

"Uniform GST is felt as the right solution to make India sustainable in Maritime Industry. The Logistic activities and formalities will be simplified", said Arunkumar Gupta, CMD of SCI in his speech in the INMARCO Conference held in Mumbai.

Trusted partners from abroad are a must for Indian Maritime Industry to grow successfully. Or else we have to be in dark with outdated systems and unskilled manpower.

Still many goods carrying vehicles do not have GPS. Hence to locate a vehicle under breakdown is a difficult task. The vehicle operator will be stranded in a remote area and struggle to inform the catastrophe he is facing.

Mobile tower will not be available in that area and lack of signals will put more hardships for him to communicate and to get in touch with his higher-ups. This set back is to be solved. A new scenario to attend to any interruption in carrying goods to the decided destination is needed.

Delay in logistics will pull down production activities and subsequently GDP growth. Good communication, clean roads and updated vehicle management systems are essential for India" says a maritime expert.

Logistic Corporation of India

"The Maritime Industry has mooted the idea of forming Logistics Corporation of India, to erase the difficulties then and there. This will assist the Supply Chain Management extensively", says Shantanau Paul of SCI.

"The cost of transporting goods in India by ship is 21% of the cost faced by road and 42% of railways. Physical inspection is 21.55% in India and it is only 6.77% in China. Multiple inspections are 83% in India, whereas it is only 2.24% in China.

Obviously, China is well ahead of us in Logistics and we have to learn from this neighboring country about the ways to handle our goods effectively, efficiently and speedily" says Paul.

Industrialists' Shift

7517 km of coastal area and 14500 km of inland water ways are not fully used as we do not have good Logistics and Supply Chain System. Freight payouts by roads are 57% and by rail it is 30%, whereas by coast water shipping it is poorly 7% of the total revenue.

This is because the maritime industry lacks skill, good representation to the Delhi rulers to expose their hardships. Budget allocation every year fails to focus high on this unnoticed sector.

As logistics fail miserably in India, the industrialists shift their production units from India to nearby countries. Delay in Logistics causes difference in production cost and it will hit badly the Indian economy, as many manufacturers prefer China.

MTO

MTO is another aspect to be seen in India with all seriousness. It is an integral part of production process. 80,000 are engaged in it and they should be experts and have financial guarantee.

MTO is a dire necessity in India, as other countries focus with intelligence and thrive. All manufacturing units are intertwined with MTO. Indian exporters need a trusted partner and this MTO is best option for small exporters in India.

Shipping Corporation of India is the first registered MTO in India.

The Logistics Supply Chain will be successful in India if we adhere to new policies and frame work for GST and MTO to make a sustainable success in the near future.

Sad state

The significance of Multimodal Transportation of Goods Act in 1993 is still to be understood clearly.

This will enable the exporters to gain support in their business growth and security in transporting their valuable goods adhering to intermodal pattern.

****

Building of New Ship Yards

Bharati Shipyard Ltd a noted company in maritime industry, has the expertise to erect multifunctional off shore support vessels, large and fast new generation OSVs for drill cutting, Tug Supply Vessels, Diving support vessels, RO-RO MPPs Container Pallet Carriers, Bitumen Tankers, Ice – Class Vessels and Jack – up Drill Rigs.

The Green field projects in Mangalore and Dabhol are note worthy. Erection of modern cantilevered type Jack – up Drill Rig on 300 acres land using largest floating docks in Dhabol has a huge fabrication hall, where assembling of materials upto 450 t can be executed easily.

2Nos of 180T cranes and 2000 T Huge Smith Hydraulic Roll Press Machine are seen here.

Ships up to 1, 50,000 DWT can be erected.

RO – RO vessels, which it builds, will possess LNG based main engines, large cargo space, etc.

Bharati Ship yard has the facility to erect PSV1400 possessing diesel electric propulsion HFO based Fuel and Catalytic Converters,150T capacity Subsea Crane, ORO capability deck area 1450 sqm, 14 independent tanks to handle mud, brine etc, and 4 tanks for Methanol.

This ship building company is specializing in defense vessels, coastal defense vessels, Navy etc. Interceptor with high speed erected by it will be a boon to Indian Coast Guard.

Bharati Ship Yard is proud to erect multipurpose off shore vessels, 150 T Bollard Pull Anchor Handling Tug – Cum offshore supply vessel and larger platform supply vessel.

Ship yard contact address:
    Dabhol:
    Bharati Ship Yard Ltd,
    Usgaon, Dabhol, Taluaka Dapoli,
    Ratnagiri 415706. India.
    Phone 91-2356248555
    Fax 91-2358 248555
    Email: dblbsl@bharatishipyard.com.

Mangalore:

Bharati Shipyard Yard ltd,
  Kuduroli, Bengere Village,
  Thanneerbhavi, Mangalore, India,
  Phone: 91 – 8242456326, 2456326.
  Fax: 91 0824456385
  Email: mglbsl@bharatishipyard.com.
  Dabhol
  Lat1 7.5868°N
  Long 73.1752°E
  Kuduroli
  Lat12.875°N
  Long 74.838°E
  Dahej
  Lat 21°42'46.55N
  Long 72°34'55.2E
  Nargol
  Lat 20°13'58.8N
  Long 72°45'0.00E
  Vansi Borsi
  Lat 20°59'27.7N
  Long 72°49 51.4E
  Kutchhigarh Dholera
  Lat 22°14'52N
  Long 72°11'4.20E
  Khambhat
  22°19'5.0 N
  72°37'8.34E
  Rewas
  23°33'49.3812"N
  75°11'43.4148"E

Aware

8°15'30.2580"N

44°19'19.4868"E

Raigad

18°44'5.5968"N

73°5'47.7672E

Vijaydurg

18°27'43.7400 N

73°50'0.8412"E

Port development is taking place in Rewas–Aware and Dighi district Raigad, Dhamankhol Bag, Jaigad, Angre, Jaigad District, Ratnagiri and in Vijaydurg, Redi District, Sindhudurg.

"Multipurpose terminal at Katale Jaigad on BOOT basis is in progress. Concession period will be for 30 years and 2yrs construction period. Techno Economic Feasibility Study Report is needed for it.

Development of support infrastructure including dredging in navigation channel and provision of navigational aids to be the responsibility of the developer, say the Port experts. Warfare charges will be as per tariffs notified by the State government.

Mandovi Dry Dock(Lat 15° 17' 57.5736", Long 74° 7' 26.3856"), a private sector engaged with 350 employees in Goa has the reputation to erect Fishing Trawler, Oil Tankers, MS Passenger vessels, Dredgers, Offshore Utility vessels, Tugs, Coastal vessels, Work boats, Ferry boats etc.

They take up complete overhauling of the above vessels and hand over it as per schedule and also at competitive rates. This dry dock can accommodate vessels upto 4500DWT and has 84mt length and 2.5mt draft.

Contact address:

Ship Yard located in Piligao, Bicholim, Goa 403504, India.

Phone: +91-832-3207436, 2363202. Fax: +91-832-2361269.

Corporate office:

Alberto Castle, 1st Floor, 27/1, Swatantra Path, Opp La-Paz hotel, Vascodagama, Goa 403802, India. Phone: + 91-83202511602, Fax: +910-852-2511596.

email:atrey@mandovidrydocks.com

****

Building of New Ports in India

The government is also interested in financing East Coast Projects at Sagar Island near Kolkata and Dugarajapattinam in A.P. Private participation in these mega projects will be an added attraction as this will enable to handle ships with 3000,000 DWT.

Apart from 12 existing ports in India, new ports in Sagar (23° 50' 19.68" N, 78° 44' 16.104" E) in West Bengal and Dugarajapattinam (17° 41' 53.999" N, 83° 16" 42.999 E) in Nellore District are to be added.

Nakkapalli (Lat 17° 24' 43.508", Long 82° 43' 45.549") in Vizag and Ramayypattinam (Lat 15° 2' 53.94"N, Long 82° 43' 45.549"E) in Prakasam are also suggested for new ports.

Ports in Dahej (Lat 21° 42' 46.561"N, Long 72° 34' 55.232"E), Nargol (Lat 20°13' 58.8" Long 72°45' 0"), Vansi Borsi (Lat 20° 59' 27.705"N, Long 72° 44' 51.452") and Kutchhigarh Dholera(Lat 22°14' 52.8"N, Long 72°11' 41.999"E), Khambhat (Lat 22°19' 5.094"N, Long 72°37' 8.344"E) are also identified suitable for ports.

Investment for Rs.8500 crores for those 6 projects is expected as on date. By these projects, the port capacity will enhance from 300 million Mt to 550 million Mt.

Mormugao Port is now engaged in construction of jetty for passenger launches / ferry. The State government is also shouldering to fulfill their aim. They are also engaged in erecting a high class cruise terminal with an investment of Rs.8.40crores. The Ministry of Tourism has offered it as a generous grant. The spade works for design and tendering are on the anvil. The port has berth for cruise vessels.

Many port projects planned by Indian government will be successful only with good road connectivity. Also suitable dredging to be highlighted or else only smaller vessels could be handled by the port. This will affect the revenue from these ports.

The Shipping Ministry of India has planned for 17 Private Public Participation Projects to the tune of Rs.10277crores. Plans like Diamond Harbor Container Terminal at Kolkata Port with an investment of Rs.1758 crores, Draught iron ore berth at Paradip Port with an investment of Rs.681crore and additional Liquid Terminal at JNPT with an investment of Rs.2496 crores are special announcements made by Ministry of Shipping.

JNPT will have a new Container Terminal with quay length of 2000mt and with a capacity of 4.8 MTEU/annum. The investment for this 4th terminal will be around Rs.7915 crores. Through Global Tender, PSA Singapore was the fortunate bidder and with this the Port Container Traffic will reach 10 million TE per year. The sophisticated technologies utilized in this terminal will positively enhance the image of this port. Automation in container handling will also be seen in that place shortly.

Jawaharlal Nehru Port plans for deepening to accommodate ships up to 16mt draught to attract 9,000 TE vessels, at present it accommodates 6,000TEU vessels with 14mt draught.

The stand alone container facility with a Quay length of 330mt with an expenditure of Rs650crores is another development in this port. Nhava Sheva (India) Gateway Terminal P ltd has taken up to this job, which is to be scheduled to finish within a year.

PSA Bharat Investments P LTD Singapore will complete 2km Berth to handle 4.8 million TEU within 2017.

Additional liquid cargo terminal to handle 7.5 million tons per year is planned. Next year the contract will be offered a leading MNC on DBFOT basis.

45 ha area of parking space is another project to be completed by next year.SEZ on Panvel – Uran road (18°58' 57.4284°N, 73°6' 9, 3636"E)is another plan with an investment of Rs4000 crores.

Solar power plant of 200kw, additional 256 reefer points, Quay cranes to handle 0.225 million TEU/year, mooring dolphins at liquid terminal, and yard for Multi Modal Logistic Park are some of the developments expected in the ensuring year.

Contact address of Mumbai office:
    JNPT, 1107 Raheja Centre, 214 FPJ Marg, Nariman Point, Mumbai – 400 021, Phone: 91-22-66165600, Fax: 91-22-67431116,
    Marketing Office:
    JNPT, Administration building,
    Sheva, Navi Mumbai – 400707.
    Phone: 022-2724 4076 / 77
    Fax: 022- 2724 4020.

Krishnapattinam Port (14° 15' 16.4952" N, 80° 6' 33.6852"E) has future projects for 7 container berths on 600 acre area. 20 Quay cranes, 40 X 42MT Rubber Tyred Gantry, with 40,000 TEUs capacity and draft of 18mt.

Port office address: P.O.Bag no 1, Muthukur, Dist Nellore, 324344, A.P India.

91-861-2377999, 91-970-41239871/989, Fax: 91-861-2377-046.

Corporate office: Plot no 379, road 10, Jubilee hills, Hyderabad 500033, Telangana India.91-40-23339992, 91-40-23557789.

While the government is interested in financing East Coast Projects at Sagar Island near Kolkata and Dugarajapattinam in Andhra Pradesh, the private participation in these mega projects will be an added attraction, as this will enable to handle ships with 3000,000 DWT. $ 550 million investment in it will boost deep sea facilities.

Essar Port is another major private player in Port industry and it has announced to clear 4 pending port projects in Vizakapattinam (Lat17° 41' 12.537"N, Long 83° 13' 6.533"E), Hazira(Lat 21° 6' 58.416"N, Long 72° 39' 6.486"E), Salaya(Lat 22° 18' 29.107"N, Long 69° 36' 2.171"E) and Paradip Port(Lat 20°16' 27.372"N, 86°30' 16.76"E) with a huge investment of Rs.3000 crores. To enhance the handling capacity to 180 MT from 77 million ton, the port authorities are investing Rs.1200 crores. 50% of the funds will be from leading banks in India.

12 more berths will be witnessed in Adani Port (Lat 21° 5' 1.029"N, Long 72° 37' 42.18"E) and government has released the relevant papers on environmental clearance. 250 MT cargo capacities in 2020 will be an encouraging scenario for this port. L & T Infrastructure Development Projects Ltd and Tata Steel have given a big support for its development with Rs.5500 crores through acquisition of 100% stake in that port.

Efforts by Insurance/Legal /Regulatory Bodies to Improve Port Sustainability in India

"The absence of competition between ports is an issue to be corrected. The policies and poor roadways are creating hindrance to profitability and sustainability of Indian ports" say the economic experts.

Healthy Condition

For Indian maritime status to attain a healthy condition all stake holders should join hands and collectively contribute to the ports' growth by having a common vision and plans.

"The Insurance Legal and Regulatory bodies should support to make the nation's blue economy to grow with 12 major ports and 48 minor ports. Good co-ordination between Port Sector and Customs Department is a must.

Apart from this, we should have quick loading and unloading, existence of mechanized operation, good night navigation, modern equipments / gadgets; less time consuming operation in container berths and nil labor problems", remarked the speakers in INMARCO-INAvation conference.

Guidelines

The Major Port Trust Act 1963 and Indian Ports Act 1908 monitor the entire activities of the Indian Ports. The Indian Constitution gives the guidelines for the safe existence and peaceful operation of ports.

The Government of India selects men on merit and experience to hold key posts and manage the Board of Trustees for all the Ports.

Safety of Ports

For minor port, the Act covers the jurisdiction of Central and State government bodies over ports. So, the rules governing operation, safety and conservation of ports are to be handled meticulously. The port dues, pilot age expenses etc, are governed by this Act 1908.

The waterfront developmental activities, plans and policies, private participation for improvement, environmental issues etc, are governed by the State Maritime Board authorities. Safety is also to be monitored by them. Many fatal disasters occur along sea-coast. The vigilance has to be stepped up. The entry of anti-national elements along Indian coast is to be eliminated.

Just in Time Policy

The absence of competition between ports is an issue to be corrected. The policies and poor roadways are creating hindrance to profitability and sustainability of Indian ports.

The long waiting time is a hurdle for huge liner ships and hence the Indian Container Cargo is obviously and unavoidably transshipped in Dubai, Singapore or Colombo. This creates additional cost and transit time. The burden falls on the poor, medium class Indian citizens, who get the pinch due to high price.

The demurrage charges, pre-berthing delays, vessel turnaround time, trans-shipment cost etc, will hit the Indian exporters and importers heavily.

This trend is not good for Indian economy, when Just in Time policy is adopted by foreign companies. We have to change their method of operation for the sustainability of Ports.

As done in Singapore

As done in Singapore, the Regulatory powers should be under the regime of Maritime Board and operation of Ports to be done by individual Port Corporation/Trust, feel the port experts.

The Port Trust will own the entire infrastructure in Port complex and rent it to private parties for commercial gains.

Restructuring Systems

The Regulatory policies and powers have to be modified and made simpler. The exporters should attain the low cost transportation advantage and manage competitions with other counterparts all over the world.

"Legal Policies should be framed by the Government to support private participation. Threats from Tax authorities and Legal bodies should be erased to improve economy. There is no point in expansion of terminals, berths and storage yards without restructuring the Insurance, Legal and Regulatory Systems", say the SCI experts.

Safety inside ports is also to be enhanced.

"Recently Jharsanya Logistics, Chennai, faced a severe loss of Rs.1.4 crores due to theft of the goods inside Chennai Port Trust. The insurance claims dragged them mercilessly and they were black-marked for immature goods handling methods", says Marshall, an executive in that organization. This company had to wind up from Chennai branch office due to heavy loss. So, the Insurance, legal and other formalities in India are not conducive to exporters and their logistic partners.

Port Acts to be remodeled

The concept of landlord of ports aided for supporting private participation in improvement of amenities, erection of container terminals with sophistication pilot age, erection of tank farms, warehouses, freight stations, dry docking / ship repair facilities, handling equipments etc.

The guidelines promoted in 1997 assisted joint ventures with ports in overseas.

The Major Port Trust Act and Minor Port Trust Act are to be remodeled such that ports flourish commercially. They can strive not only to safeguard their turfs but also to give a financial nod even to an urgent project to improve business.

"This prevents the importers and exporters to face competition. Replacement of cranes, spare parts procurement, warehouse modifications and extension plans are to be done with approval from union ministry. The powers to sanction bill more than Rs.5 billion is needed now for Port Trusts", viewed the Shipping & Maritime Expo conference speakers.

The Chartered Accountants employed there have age old accounting practices and it should switch over to commercially.

"Corporatization of Ports will remove many bottlenecks to raise funds and liberal privatization" says an economist.

TAMP

TAMP is another milestone for removing bottlenecks. It should be fully utilized and made powerful enough to monitor regulatory activities. Apart from Tariffs steps for port improvement is a must. Fixation of Tariff based on private participation, to improve efficiency and operational standards are to be done. Obviously they increase or decrease tariff, which is not encouraging in port sector.

Each port has different tariff and accounting pattern. This confuses the importers and exporters. The foreign ships entering Indian ports consider it as a big bottleneck for port economy.

Single body to fix Tariff, to manage accounts and to dexterously handle up-gradation of all port activities is needed in India. Then only we can see a healthy competition among the ports.

Just like river disputes and bottlenecks in river water sharing are solved by Tribunal, the port scenario also needs a Tribunal to solve any disputes, which cannot be solved by TAMP.

As private participation increases, the existence of TAMP will be questioned, say the port experts.

Clash

Another major bottleneck is that the Port Trust authorities and private investors, clash in administration activities.

The private players have a tug-off war with Port Trust with regard to berth facilities, as new terminals and berths created by them are inside the same Port Complex, where Port Trust enjoys certain berths and infrastructure facilities.

Monopoly of private investors will raise its ugly head if no competitor is allowed to establish container handling facilities.

Quickness Needed

Lack of road connectivity kills competition between the Ports. So, to eliminate this bottle neck, sufficient funds for roads, land acquisition, simpler formalities for legal handing over of land documents by land owners, payment of assessed value imminently and clearing road project tenders are needed.

Labor Problems

As most of the goods in ports are handled manually, the workers play a major role in port economy.

Quickness of loading and unloading activities in warehouse should be modernized. Or else the Dock Workers Act 1948 will be a big headache to private investor. The Tariffs are to be modified for favoring the private players in port.

Good understanding between private investors, ware house authorities, Insurance agents, Safety Officers, Regulatory Authorities, Environmental exports etc. is needed always.

Radical Changes for True Sustainability

Landlord concept enables the port authority to have full control on infrastructure and land and allow the third party to execute operations in that port. A radical change in legal and institutional set ups are needed in India.

The roles, responsibilities and obligations of public and private players should be clearly indicated in master plan and vision. Policy to correct Dock Labor laws is to be attended.

Environment, safety, investment etc. are to be planned.

Legal aspects for land acquisition, development, project implementation, contract complications, arbitration formalities and legislation for private players to play a vital role in port sector are needed.

Encouraging private parties with incentives, subsidies, liberal loans, investor friendly packages, elimination of enormous formalities to clear the documents and quick disposal of environmental clearance papers are essential for improve sustainability.

Automation, computerization and quick material handling of the goods to be exported and imported play an important role in Indian Port Economy.

Dredging of channels with overseas experts is a must.

Then only we can dream of true sustainability in Indian Ports.

****

Coastal and Inland Waterways

"Seamless operations between RSV/IV and Coastal water ways needed" say the economists.

The formation of Jal Marg Vikas is a big step for success in Indian water ways with the support of World Bank assistance of $ 700 million.

Depth increased

Inland water way 1 to 5 plays a leading role in Indian economy.

The present trend of utilizing National waterways for trade by cargo vessels, tourist vessels, ODC carriers and IWAI vessels is a matter of appreciation by economists as it improves GDP. NTPC is receiving coal from Haldia up to 3MMTA availing the National Waterway 1. When power crisis hits many industries, the movement of coal is praise worthy gesture in Indian rivers like Ganga – Bhagirathi and Hooghly River for a distance of 1620 km.

Even though least available depth is 3.0mt only, 2.6 lakh tones have been transported effectively. The fertilizer movement up to 2600T, by Tata Chemicals is also another landmark in water transportation. Inland water way has also boosted tourism through vessels like RV Bengal Ganga, ABN Sukhapa etc. The depth has been increased from 2.0mt to 2.5mt along Farakka Buxar. Again another step is taken to deepen by 2mt up to Varanasi.

Usage by BSF

The formation of Jal Marg Vikas is a big step for success in Indian water ways with the support of World Bank assistance of $ 700 million.

National water way on the river Brahmaputra for a distance of 891km has aided GDP growth. The usage of water ways between B'Border Neamati with 2.5mt depth and between Dibrugarh – Sadiya with a depth of 1.5mt by Border Security Forces is the highlight of IWAI. ODC and lengthy cruise vessels operate in NW2.

As it covers Manas Wild Life Sanctuary in Jogighopa, the wild animal lovers wish to use this water way. It is a great advantage for Indian Merchants that even Petroleum, Oil and Lubricants are transported through this low level water path between Silighat to Budg – Budg and Banghmari. Cutter suction dredger is the correct solution to deepen the shallow areas in Pandu – Neamati Coastal area. Solar operated Navigational Light support the movement of foreign tourists along B'Border to Dibrugarh. More than 125 to 265 Navigational lights are needed for safety in this water way.

Needs of IWT Fleet

Cargo vessels and passenger ferries are to be operated at Pandu, Tejpur, Togighopa, Dhubri, Neamati, Silghat, Bogibil, Dibrugarh, Biswnath Ghat, and Sengajan Oriumghat. For these projects desired activities like land acquisition, shore development activities erection of crane pontoon works have been done.

Sengajan is a vital location, as floating terminal is maintained for the sake of Indian Army Vessels between Arunachal Pradesh and Assam. Defense men watch this waterway to fulfill the needs of IWT fleet of Indian army.

BSF also utilize Oriumghat floating steel pontoon. 11 locations are utilized for setting up terminals in National waterways – 3.

Coal and Rice Transported

Alappuzha, Thirukannapuzha, Kayamkulam, Kollam, Aluva, Maradu, Kottapuram etc. are some of the important locations in NW3. We have storage facilities ranging from 200sq mt to 400sq mt open storage.

To handle the goods 18T mobile crane and 3T capacity fork- lift are installed in all locations. Approach road for these areas are well maintained and 30m RCC jetty is seen in all locations.

NW4 covers Godavari & Krishna Rivers & Canals stretching between Kakinada and Pondy.

Coal, Cement, Rice are transported by this useful waterway. Brahmani River and Mahanadi delta system form the NW5.

Seamless Navigation

Directorate General of Shipping aims for Seamless Navigation between inland water and the sea. It will safely evade the Coastal Shipping Act. The notification of River– Sea Vessels gives significance to stability and safety of these vessels travelling on shallow water. The new Inland Vessel Act 2008 enhances the commercial gain in this poor country as it paves way for the operation of IV vessels from one port to another port and shore to ship operations during day time and during fair weather condition. This is to be covered under type 1 & 2 of inland vessels respectively. The vessels types 3 & 4 have been covered under newly framed IV Act to operate in fair weather condition round the clock. The type 4 vessels will be covered for operations under all weather conditions.

The River–Sea notification restricts the cargo carrying capacity up to 3000 gross tonnage only. So, by this law, the vessels carrying this limited weight cannot load higher volume of goods when it reaches sea area, where there is restriction of draft.

The Maritime experts raise the question "How the River-Sea notification will improve trade, when it has restricted the load upto 500 tones as cargo carrying capacity?"

The idea of seamless integration of sea and river is not economically healthy. We have to seek the multimodal route from river to coast and vice versa, say the water experts.

Obviously a shipper will ship his cargo based on cost of transportation from origin to destination. In this context we can say that River-Sea-Vessel notification will not fetch satisfactory and conducive atmosphere in India. Through IV Act, we have to facilitate port to port and ship to shore movement in Indian water.

Coastal Liner Service

The act of SCI to organize and operate Coastal Liner Service is to be appreciated. The liner service from Karachi to Chittagong to operate along all ports is a heartening one. The Coasting Vessel Act is ignored and SCI is operating service all along the Indian Coastal Area.

The Merchant Shipping Act 1958 on Coastal Operating Vessels has been fine tuned by RSV 2013 to enable pollution free and accident free travel. Importance has been given to better fire fighting system, radio, communications, carriage of cargo etc.

GDP Growth

The RSV notification aim for better GDP growth due to hassle free and rich coastal trade in India. The pitfalls in Merchant Shipping Legislation, which caused coastal shipping expensive, have been eliminated by RSV 2013. The cost of raw materials and exorbitant labor cost has shot up the constructing and operating expenses. This creates headache to coastal and inland shipping trade activities. The River sea vessel notification has gained importance in India as it gave due importance to safety of coastal ships wrecks, prevention of collisions of ships etc and thus revenue loss in trade have been drastically removed.

Cheered

MSA amendments in RSV 2013 have been cheered by all entrepreneurs in Indian Coastal business. The River-Sea-Vessels' trade between India and Bangladesh with 2.5 mt draft with great fuel savings is appreciated all over the world in Maritime Industry. The design of RSV-4 ships will suit to any weather scenario and seamlessly operate from sea into land water ways and vice versa.

Insurance companies in India will be benefitted as all the coastal vessels insured in Bangladesh will also be insured and registered with Indian authorities.

Maritime sector with its new policies will pave way for a rich Indian economy in the future.

****

Inland Waterways in India

Inland waterways are important for economic growth of a nation. Hence importance should be given to it.

Six new National Inland Water Ways Transport Plans is yet to be sanctioned by it. Hence Tamil Nadu and Pondicherry is yet to get the benefits of Kakinada – Pondicherry inland water way transport system.

For the past 10 years, inland water transport system is not utilized properly, even though there 6 major rivers in our country! At present, in the rivers stretching for length of 5200 Kms and in the canals stretching to a distance of 4000 Kms, with a total distance of 14500 km length of water ways, all ships and boats sail and carryout the inland water way transport system in India.

When compared to USA, European Countries and China, it is very less here. In USA, 21% of goods are transported by inland waterways only. But in India, 0.1% of the goods utilize inland water ways for transportation, which is very poor. So, to enhance this transport system, Indian Inland Water Way Sector functions in Noida (Uttar Pradesh). Its divisional offices are located in Patna, Kolkata, Guwhati and Cochin. Sub-divisional Offices are in Allahabad, Varanasi, Bhagalpur, Quilon and Barakka.

The Indian Inland Water Transport Division authorities chalked out 6 new inland water ways.

This plan will utilize the rivers, canals; water drains etc. for transportation of goods for economic development of all the States.

National Waterways 1: This is 620 Km stretching between Ganges, Bhagirathi and Hooghly rivers located in the area between Allahabad and Halda.

National Waterways 2: It stretches to a distance of 821 Km in Brahmaputra between Chadia and Dhubri.

National Waterways 3: It is between Kottapuram and Quilon covering the canals like west seashore canals, Sambakara Canals and Uthyogamandal canals for a distance of 205 Km.

National Waterways 4: This water route utilizes Kazhuveli Kulam between Kakinada and Pondicherry; Godavari River between Rajamundhry and Badrachalam and Krishna River running between Vasirabad and Vijayawada. Total distance of this inland water ways is 1095 Km.

National Waterways 5: Brahmani River running between Thalaicherry and Thamra; Canal between Geonkali and Charbadia along East Sea Coast; Madai river running between Charbadia and Thamra; Mahanadhi Delta covering Mankalkadi and Bharadeep, are under this water way. Total distance is 623 Kms.

National Waterways 6: Bharath River between Lakhipur and Bhanka covering a distance of 121 Km is under this Water Way Plan.

For Tamilnadu, the 4th Inland Water Way is important. Through this plan the rivers Godavari and Krishna will be linked by a canal from Kakinada to Pondicherry. Total distance is 1027 Km. River portion is 328 Km, canal distance is 302 Kms and Salt water canal portion is 397 Km and the total estimate is Rs.1515 Crores.

For the first phase Rs.609 Crores and for the second phase Rs.906 Crores are to be spent. To erect West and South Buckingham Canal, Kammavoor Canal and Kazhuvelikulam Canal the estimate is Rs.906 crore. It is planned to finish it by 7 years. To execute this plan, 300 hectares in Tamil Nadu, 380 hectares in Andhra Pradesh and 27 hectares in Pondicherry had to be acquired.

For this land acquisition Rs. 391 crores is needed and for canal construction Rs. 333 crore is needed. Utilizing this Inland Water Way Plan, we can transport Lignite along river Godavari, cement along the river Krishna and Rice via Krishna and Godavari. Totally Rs.10 Lakh worth of goods can be transported by these water ways.

Utilizing the Buckingham Canal, water ways transportation of goods can take place to Andhra, Tamil Nadu and Pondicherry. Many industrial developments can be witnessed due to it.

Ship Transport:

Instead of transporting goods via Lorries, we can use ships to transport, by which we can save 90% of petrol and diesel consumption. Lignite and coal can be transported in large quantities to generate power in these states, where these fuels are needed. We can save Rs. 65 crore worth of oil and reduce unwanted import at high cost. By these inland water ways, floods can be avoided as rivers are linked. Unnecessary expenditures and flood relief can be avoided. Many crores lost in industrial production due to floods can be avoided.

Drinking Water Supply:

Using the inland waterways, drinking water can be supplied to the needy areas.

Irrigation:

Due to this great project, 15 crore acres of land can get irrigation and produce 50 crore tones of food stuffs.

We can avoid unwanted foreign exchange also.

Power Generation:

Due to inland water ways, 60,000 mw power generation is possible to fetch an income of Rs.60, 000 crore. The Central Government has given green signal for it and has included in 11th Five Year Plan also.

The Planning Commission should allot funds for Kakinada – Pondicherry Water Way Project says the water way experts.

Buckingham Canal was mainly erected to give employment and to face famine.

The length of the canal is 426 K. It covers 250 Km in Andhra Pradesh and 170 Km in Tamil Nadu. This lengthy canal links Mamallapuram, Thiruporur, Tiruvanmiyur, Chennai, Pazhaverkadu, Vellore, Ongole, Masulipattinam, Vijayawada, Kakinada, Polaravaram and Vasirabad. Rice, Vegetables, Milk, Sugar, Fish, Mutton, Firewood and Chicken were transported earlier. Due to improper maintenance, these activities were cut for the past 50 years. Along these canals, various hotels, industries and houses were erected.

Their wastes spoiled the canal and made it an unusable drainage. Good water way is now stinking. If the government allocates funds, then this canal will get uplift and restore Inland Water Transport to improve the industrial scenario in the two States. The people living along that area will also be free from contagious diseases like Malaria, Dengue fever, Diarrhea and skin diseases.

Shipping & Maritime Expo 2014 Conference

'Foreign entrepreneurs are very much interested to invest and start their automobile industry in India. But no major ship building company is willing to start their unit in India. So, no improvement is seen in our maritime industry" sadly exposed Dhowan, Chief of Naval Staff.

Marine Experts

The Shipping &Maritime Expo 2014 INMARCO-INAvation 2014 was organized from Dec 11th to 13th 2014 by The Institute of Marine Engineers (I) Mumbai Branch, Institution of Naval Architects and CII in a grand fashion in NCPA Complex, Mumbai.

IR Class had the honor to be the principal supporters of this intellectual and interesting meet.

The organizing committee was studded with Marine experts, Port Chiefs and Ship Yard VIPs like V.Kumar, Chairman of Bharati Ship Yard, Saxena Chairman CII and Maritime industry experts like V.K.Jain, S.M.Rai and Umesh Grover.

The Advisory Board of the great conference had Atul Agarwal, Managing Director of Mercator ltd, R.M. Pamar IAS and Chairman of Mumbai Port Trust, Inspector General S.P.S. Basra, Cost Guard Region (W), Vice Admiral A.V.Subedar, Director General Naval Projects Indian Navy, A.Banerjee, Ex.Chief Surveyor Director General of Shipping and many more.

Prabhakar Srivatsava, Chairman Mumbai branch IME welcomed the gathering. Vice Admiral A.G.Thapliyal AVSM and BAR , Director General Coast Guard, C.V.Subbarao, President IME and A.K.Gupta, Chairman CII addressed the audience of maritime delegates from all over the world.

Chief guest Admiral R.K. Dhowan, Chief of Naval Staff inaugurated the show.

Competitive Sea Farers

Noboru Ueda, Chairman Class NK, Japan stressed that India needs competitive seafarers and talented professionals for innovations. He assured to assist Indian maritime industry by providing complete knowledge on ship and ship building, inspection services and procedures.

"Bright Indian professionals are a must for the present downtrodden condition of the shipping sector" said he.

C.V.Subbarao, Chairman IME, India said there are plans to acquire 24 more vessels and 12 offshore vessels. Good propulsion systems, new paints to erase marine pollution and fire fighting sensors are needed for Indian maritime to grow. Support by technologically sound man power is vital, said he.

Lack of Facilities

Dhowan, Chief of Naval Staff openly disclosed that "Indian shipping industry needs revitalization. National security advisors have also warned about it. Issues like carbon release, service tax and lack of facilities make the foreigners and MNC to refrain from buying ships in India. Ship building activities are hindered by trade employment, FDI and foreign exchange issues.

"Files in defense deals had to be settled by another division in Delhi. This creates delay for approval. So, one single team is needed to clear the shipping and defense proposals. Good vision, goal and implementation with dexterity are needed in India. Foreign entrepreneurs are very much interested to invest and start their automobile industry in India. But no major ship building company is willing to start their unit in India. So, no improvement is seen in our maritime industry.

Automatic navigation will save $ 1.3 trillion, says Morgan Stanley report. Importance of SAM strategy is to be highlighted in Indian maritime sector. Train more crew digitally and use ship shore system to de-skill on board tasks" said he.

Major sea areas untouched

Good inland water way connectivity and utilization of wind energy to produce 74 000 million megawatt are needed now. 7% of world sea farmers are Indians and 150maritime institutes will positively increase it to 9%.

90% of world trade is done by sea. But only 8% to 10% is carried out in Indian sea. This is a sad situation prevailing unnoticed for a long period.

Major sea areas are untouched by fishermen. Deep sea fishing is the solution for creating jobs.

Only 11% of our oil requirement is obtained from Mumbai and Krishna regions. Many more blocks are to be demarcated for exploration.

Terrorists enter easily

Indian sea coast is no longer safe and maritime area is extremely dense. 46 Coastal Radar Systems, Auto Identification System, Information Management System, 31 Coastal Guard Stations, 20 Naval Security Guards and 9 Coastal Police Stations are needed. The staff employed should be able to speak the local language fluently and then only the Indian Coast will be safe.

Or else terrorists will make an entry easily.

Support from local folks along the coast is not encouraging to identify the intruders.

It is disappointing that 70% of national disasters occur on Indian Ocean.

Increasing Indigenization

Indian Naval force gives a helping hand at times of Tsunami, floods in Jammu & Kashmir, and Andhra Pradesh. 41 ships and few submarines are under construction. All are done in Indian ship yard only.

At Kochi, erection of ship for air craft carrier is in progress. Indigenization of auxiliary machineries has to be enhanced significantly. Even though we have our own radars and missiles, the Indian private industries should involve in increasing indigenization. Training abroad for Indian technicians is needed".

He proved to be an eloquent speaker with various statistics in his finger tips.

Jt.Director General of Shipping, Deepak Shetty expressed that energy security and transport security to be modernized and upgraded in India.

CFD and Dual Fuel Technology

Sriramamurthy, COO, IRS viewed the usage of CFD technology for innovation in this under developed country.

The speakers raised the issues due to hike in fuel cost and remarked critically that cost of fuel is as much as ship building cost in India. Weak customer relationship, shortage of knowledgeable crew and slim technical resources are causing setbacks in India, said a speaker. Digital revolution, good team work and orderly organization structure are needed for the hour.

Dual Fuel Technology to extend tank capacity of liquid fuel and gas as backup fuel to eliminate cost of fuel and to control emission, etc were discussed at length in the conference by Das Gupta. 2000kct of under developed gas is still remaining untouched and 40% gas is stranded in India, said he.

Usage of scrubber for eliminating Sox faces problems like heavy vibrations, corrosion in piping, increased back pressure and heavy cost, which have made the Indian ship builders evade it. The cost of it is $2 million US dollars. If sea water is filtered and reused for drinking water, it can be economical for Indian ships.

If 0.5% sulphur content regulation is implemented, then more Indian ships will have scrubber by 2018, said Dr. R.Vis Visweswaran.

Green Technology

Green technology for sustainable shipping using variable frequency device was explained by Prof Dabadgaonkar. Usage of fuel will be controlled by Variable Frequency Drives (VFD) and more Indian ships have to utilize it to reduce emission issues, said he. "For 30% reduction speed of motor, we achieve 70% lower rpm and hence power consumption will be only 34% and saves fuel. 5% speed reduction will save 13% fuel for tankers and 19% for container ships", expressed Dabadgaonkar.

Regulations governing shipping industry are increasing and environmental based economics will affect the maritime growth in India. Finding an alternate fuel and exhaust gas treatment technique was also discussed in the conference. Integrated Transport Policy is needed to strengthen Port, said the speakers.

Flaws

Uniform GST (Goods & Service Tax) will give support to Indian maritime industry, said Shantanau, GM of SCI. "Lack of ECDIS knowledge, ignorance in equipment standardization, ineffective implementation and non-assessment of time frame cause flaws in Indian Maritime sector. Confidence to ask for help, to understand and operate, minimum standard for training, reliance on technology with checks and realistic assessment are needed", said he.

****

Shipping & Maritime Expo 2014

Shipping & Maritime Expo 2014 took place in NCPA complex .

Totally there were 27 participants.

Delegates and visitors in thousands enjoyed the show on all the 3 days. Especially the international traders, who were interested to have tie-ups and collaborations, were seen busy in discussions for further business developments. The Marine Institute students had a good time to observe the latest trends and innovations in maritime sector.

On the whole it was a useful exposure to all participants.

Bharati Shipyard, the leading ship builder in India took a good opportunity to reveal their greatness in this field, through this informative INMARCO-INAvation 2014 Expo 2014, held at NCPA Complex in Narimanpoint, Mumbai.

This company listed in Bombay Stock Exchange and NSE has won laurels in taking up ship erection works in Mumbai, Kolkata, Goa, Ratnagiri, Mangalore and Dabhol.

In Mumbai, it is engaged in steel fabrication and major piping works on 12 acre land with CNC Plasma cutting, Panel line fabrication and Auto welding machines. It is located in Ghodbunder.

Bharati Ship Yard Ltd has the capacity to build hulls up to 125mt length in Kolkata, vessels up to 150mt length in Goa, vessels up to 220mt length in Mangalore and in Dabhol Yard. It can erect Jack up – Rigs, Off-shore structures and vessels up to 100,000 tons and 220mt length. Dabhol Yard is the largest among all with an area of 30 acres.

They have been asked to erect 150T DP – 2 AHTVSVs, 220t Diesel Electric DP-2 AHTSVs, 100m Diesel Electric DP-3 MSVs, 80t DP-2 AHTSVs, 4000T DWT DP-1 MPSVs and 3300TDWT DP-2 MPSVs.

With 3 level quarters for 110 individuals, Heliport, 3Nos of PCM-120ss cranes, 60 nos of 375 –Kip Electro mechanical Rack and Pinion Drives, Letourneau Super 116E class platform 65T,80T to

120T, Bollard Pull Fire Fighting, Anchor Handling Tug Cum Supply Vessels etc were all erected by this reputed ship building company. Diving Support Vessel, Offshore Crew Boat and 100 m Tanker cum RORO vessels were all fabricated and erected by this company.

Contact address: Bharati Ship Yard Ltd,
302, Wakefield House, Sprott Road,
Ballard Estate, Mumbai – 400 038. India.
Phone: +91-2230289200.Fax: +91-2230289222.

Saifee Enterprises, which imports aircraft tools, power tools, offshore tools, wood working tools, automobile tools and industrial cutting tools, had occupied a space in this expo. Contact person: Hunaid Lokdandwala. Email id: saifee.ent@gmail.com, Cell +91-9867645115.

Contact address: Saifee Enterprises,
44, Nagdevi, X lane, 2nd floor,
Office No.20, Mumbai, 400 003, India.

GOL Offshore has its office in Energy House, 81, Dr.D.N. Road, Mumbai-400 001. Phone: 91-22-66352222, email:bdmoff@goloffshore.com. They exposed their various activities in this great exhibition. For the past 32 years they have been engaged in marine logistics, drilling, marine and offshore engineering ship repair maintenance, port and terminal support etc.

Not only in India, they have served in middle east with 4 offshore construction support infrastructure, 2 construction of Hook up Barges and construction support for Work Boats. They have served with PSV, AHTSV, OSV and AHT in JNPT, Tutucorin Port, Kolkata Port and private ports like Gujarat Adani Port, Gangavaram Port and Pipava Port etc.

"Rescue Operations done by them for Indian Naval Frigate, Product Tanker in Mumbai Port and General Cargo vessels erected by them are lauded by all", said their stall staff in this show.

In Mumbai it has its own 2,400mt floating Dry dock.

TEBMA Ship Yard Ltd, Malpe Harbor Complex, Malpe, Udupi, 576106, Phone:+91-8202538900 occupied a space in this show to exhibit their potentials in erecting Multipurpose Support Vessels, Cutter Suction Dredgers, Utility Vessels, Geo Technical Research Vessels, Harbor Tugs and Special Purpose Customized Vessels. Their corporate office is in 40, Bazullah Ro, 2nd floor, M.T.Rajen Properties, T Nagar, Chennai – 600 017, T Nadu. Phone: 91-44-28340703, Fax: 91-44-28340702.   Email: info@tebma.com.   Email: info@bharatishipyard.com.

An agency is always needed to represent the grievances of, Ship owners, operators in Logistics, Manufacturers of Dredgers and Offshore Support Vessels, companies involved in Cargo Ships, Port and Terminal Operators. So, ICC shipping Association does it to solve issues with the government departments, environmental problems and tax concessions. They had a stall to explain all the above activities. Contact person: Jyoti Birare. Address: ICC Shipping Association, Scindia House, Basement, N.M.Marg, Ballard Estate, Mumbai 400001,   Phone.022-22623003,   22623912.Fax: 022 22623911,Email:mail@iccsa.in.

Chugoku Paints (1) P Ltd with their office in 405, Raheja Chambers, Free Press Journal Marg, Nariman Points, Mumbai – 400 021, Phone:022 - 43550600, Fax: 022 - 43550625, email: sales@cmpindia.net. Environment friendly, low labor expenses, longevity and fuel saving are some of the special features of their Seaflo neo, said the representative in the stall allocated for this company. Ramesh Tripathi, their Regional manager can be contacted by his mail id: rtripathi@cmpindia.net. Mobile: 09819943312.

Taurus Marine Services (P) Ltd, 402, Vidarbh Vaibhav, 64, Bhavanishankar Road, Dadar (W), Mumbai-400028. Phone: 91-22-24391700, Fax: + 91-22-24391719, email:tms@taurus-marine.com showed the Induction Buffet Warmer, 100Lts Kadai, Bulk Cooking and Fulka Puffer. They say that 16kwh Induction Heater will save Rs499800 per year and recommended for ships to save fuel and cost instead of using LPG burner.

Taurus Marine Services take up replacement of worn out marine parts in ship.

F.O., D.O & LO purifiers, Centrifugal Pumps, Flow Meters, Sludge Pumps, all types of Compressors etc are handled by them.

Mrs. Anjali Bhide is the Managing Director of this esteemed company, which has won ISO 9001; 2000 DNV certificate.

It seeks international customers for support in all service activities needed for marine industry.

Aryatech Marine & Offshore Services P Ltd, B-1, Hauzkhas New Delhi – 110016, Mob: 91-9999009126, Phone: +91-11-46018102, Fax: +91-11-46018103, Email:info@aryatech.net, was there in the show to tell the public about SACS (Structural Analysis Computer System), a product of Bently Systems. USA, Auto plant, Multiform, Auto pipe, STAAD PRO etc are some of the software they deal.

Indian Navy utilizes Max Surf to erect floating structures, calculate stability etc.

They render training in Subsea Systems Mooring, O&M of Offshore structures etc.

Naval Architecture Drill Ships, Semi-submersibles and Single Point Mooring are some of the areas they serve.

Essar Trading in Mumbai supply Piston type valve, Gate valve, Check valve etc. Pressure ranging from class 150 to class 2500, various sizes from 15mm to 250mm are marketed by them.

They are manufacturers and suppliers of Eccentric Wheel Assembly 200m to 260mm with all spare parts regarding it. Rescue Boat Embarkation Light, Inclinometer, Electrical Cables, Plastic Cable Tie, Air Horn, Emergency Light etc are some of the products they deal with. Their office is located in Essar Trading & Engg Co,44, Nagdevi Cross Lane, 2nd floor Room 20, Mumbai -400 003.

Contact Person: Ravi Mehta, Jivan More. Mobile Phone: 9967556022, Land Line: +91-022 23420484, 022 22136537.

Ken Mark Tech solutions with branches in UAE, New Delhi, Goa, Surat, Bangalore and Chennai, has its registered office in Mumbai.

They render services for Oil pumps, Air compressors, Emergency Engines, Heaters, Boiler and Turbine set for power generation from 0.5mw to 2000mw.

Repair works in cracked engine casings, metal stitching in engine cylinder head, reconditioning of any part of the ship etc are carried out by them.

Timely delivery competitive rates and talented technical team are plus points for this company, said their representative.

Their workshop is located in A-3/A-4, Singh (1) Industrial Premises Co-op Society, Ram Mandir Road, Goregaon (W), Mumbai, 400 104. India. Email: info@kenmark.in.

Aries Technical Service P Ltd is located in D-2, Udyog Sadan-3, Maroli M.I.D.C, Andheri (E), Mumbai, 400093. Phone: +91-2240271800, Fax: + 91-22-400271818, Email: info@ariestech.co.in. They deal with clutches manufactured by Quick Shift USA. They are interested in dealing with other marine equipments, sales and service with accuracy.

NUSI Offshore Training Institute had a stall to tell the visitors about the training they impart to serve to Maritime Industry efficiently with good DP Simulator, Crane Simulator, ROV Simulator, OSV and Engine Simulator. The students gain abundant knowledge practically to face any problems on ship. Excellent Cafeteria, Gymnasium and green garden make the students study with all relaxation and joy. Their office is located in NUSI Offshore Training Institute, Thakurwadi Road, Wardoli Village, old Mumbai, Pune Highway off Shedung Phata, Panvel, 401206. Email: nofipanvel@gmail.com.

Class NK, which deals with safety, environment, education, training and prime management had a spacious in this expo to tell the visitors about their services and products.

Software dealing with operation, maintenance and machineries of ship are their valuable services.

It has a worldwide service network in China, Singapore, Dubai, London, New York and Rio de Janeiro. They are yet to open their office in Germany. Their service centres are in Algeria, Amba, Argentina, Australia, Bangladesh, Brazil, Canada, Dalian, Colombo etc and many more countries like Yemen, West Indies, New Orleans, Los Angles, Seattle, Ukraine, Uruguay, etc.

The exhibition conducted along the sea shore of Mumbai was informative and interesting.

More than 15000 visitors and delegates from all over the world made it a grand success.

****

Development of minor Ports in Tamilnadu!

With private participation it has been decided to develop the minor Ports in Tamilnadu to enhance industrial and economical growth in Tamil Nadu.

This step is to be effectively implemented with liberal funds and efforts.

In Tamilnadu we have 3 major Ports like Chennai, Tuticorin and Ennore Port.

Apart from these Ports we have 15 minor Ports like Cuddalore, Nagapattinam, Rameswaram, Pamban, Kulachal, Valinokkam, Kanyakumari, Ennore, Punnaikayal, Thirukadaiyur, Py3, Kattupalli, Thiruchopuram, Manapadu and Kudankulam.

Considering the gains obtained from development of these Ports, the Ministry for Seaport is taking active steps with huge funds. By developing these minor Ports, we can expect growth in Thermal Power Stations, industries based on production of anchors, chains and sea foods. Oil Industries will also thrive with their support says the experts. In 2006 there were 15 minor Ports in Tamilnadu and it has increased to 20.

Cuddalore Port is at a distance of 200km from Chennai. Tamil Nadu Government planned to erect walls to avoid erosion, new administrative buildings and new wharfs on 100 acre area. Due to storage of funds the works were held up. So it was decided to handover the projects to private parties for development and maintenance for a 30 period of lease.

Chemplast, a group India Cements has created a Marine Terminal facility in Cuddalore to handle their Polyvenyl Chloride Chemicals.

Cuddalore Powergen Corporation has planned to erect storage facilities in Cuddalore port at an estimate of Rs.325 crores.

Chennai Petroleum Corporation has taken up Oil refinery plant erection and oil handling facilities in Nagapattinam Port.

Such participation of private agencies only can enable a sea-change in Port's infrastructure in Tamilnadu.

Present condition of Minor Ports

Kattupalli: -

This is near Ennore.

At an estimate of Rs 3375 crores ship building facilities and private Port erection are in full swing.

Mugaiyur: -

This is near Mamallapuram.

A small Port erection was announced by the Tamilnadu Government. A ship building facilities with Rs.500 crore estimate will be witnessed shortly.

Thiruchopuram: -

This is located in Cuddalore district.

Nagarjuna group has been permitted to erect a Port with an estimate of Rs.800 crore. Crude Oil and Petroleum will be imported and refinery works will be taken up by them.

Silambimangalam: -

Good Earth Ship Building a Private Company has been sanctioned to erect a Port here with an estimate of Rs.500 crores.

This is near Cuddalore District.

Ships weighing 75000T can be anchored in this Port.

Kaveri: -

Bell Power, a multicrore private company has planned to erect a 320 mw power plant in Poompuhar in Nagapattinam District.

To handle the required coal for this power plan, ultra modern facilities will be erected there by it.

This will give a big uplift for Kaveri Port.

Vanagiri:

This is located in Seerkazhi in Nagapattinam district, where a 500 mw Power Plant with an estimate of Rs.250 crore will be erected by NSL Power.

A big facelift to that Port will be given to handle the coal required for it.

Thirukadaiyur: -

This is also located in Nagapattinam district.

In Pillai Perumalnallur, PPN Power a private power generating company has started a 330mw Power generating plant.

To handle Naptha and natural gas for this plant Thirukadaiyur Port was erected in 1996.

Thirukuvalai: -

It is located in Vettaikaran Iruppu in Nagapattinam.

Tridem Port and Power Company has erected 2000mw power plant at a cost of Rs.650 crores.

To handle coal for this plant a Port has been permitted to erect on 276 acre land on lease.

Manapadu: -

This is located in Tuticorin and it was erected at a cost of Rs.800 crore.

Indian Power Project has planned to erect a 200mw gas based power plant near it and 100 acres of land has also been allocated for this huge project.

The Port will be supporting the needs of that power plant.

Kudankulam: -

Kudankulam Atomic Power Plant is located in Kudankulam (Tinneveli District).

For this power plant, arrangements to store sea water for a length of 1km and width 500mt are being executed.

With an estimate of Rs.340 crores Jetties and other amenities are being erected.

Panayur: -

In this Port a facelift is being provided to handle lignite and coal for 4000mw power plant to be erected by Coastal Tamilnadu Power Limited.

Parangipettai: -

This is developed by I.L. and FS Ltd to cater the needs of 4000mw power plant which is to be erected at an estimate of Rs.300 crore. Coal handling facilities will be provided here with all sophisticated equipments.

Udangudi:-

Chennai Udangudi Power Corporation Ltd has planned to provide facilities like wharfs, cranes etc for unloading goods from ships.

Also amenities to support 1600mw Super Thermal Station which is to be erected here will be provided.

Kasimedu Fishing Port

"Due to lack of maintenance, the Chennai Kasimedu fish Port will collapse within a year or two" say the specialists.

It is high time the authorities take some smart steps to protect it.

For all fishermen in Chennai, Tiruvallur and Kanchi District, this Kasimedu Fish Port has been an important Port to fish for their livelihood.

There more than 1300 motorized boats and hundreds of fiber boats are engaged in fishing. Around 30,000 men are engaged in this profession. Out of this 13000 are involved directly in fishing.

Indian earns a foreign exchange of Rs.10500 crores through export of fish and out of which Rs.500 crores is earned from Tamilnadu.

The fish export business has played a leading role in Tamilnadu's income.

When such is the situation it is surprising that State Government and Central Government ignores it with least care.

At times of election they speak high about budget allocation to develop fishing and fish export activities.

Several crores are also available but nothing is done usefully or fruitfully. So the age old fish Ports are still in uncared condition which is totally unhealthy for economic growth and also for environment.

Big holes and pits are seen in wharfs and 3 fishermen have slipped and fallen into it.

Yet no repair work is taken up to set right it.

Some temporary supports are seen which are not sufficient for safety and security.

Many areas where fish are auctioned are in unhygienic and filthy condition. Maintenance and upkeep are pitiably ignored. Even rickshaws cannot enter to carry diesel for the motor boats.

Basic amenities like lighting, toilet, etc are neglected. Even a decent restaurant could not be seen for fishermen to have a healthy food.

When crores are allocated for betterment works it is unbearable to see that area filled with pungent smell and rubbish.

Huge money is handled here and no government or private Bank is seen there to handle it and safeguard it. After each fish trading transaction is over it is unsafe to carry that lump sum in gunny bags all the way to a Bank in a distant location.

Due to improper lighting the diesel thefts are common here.

Sometimes boats are also stolen by miscreants.

The Tamilnadu government has handed over Rs.20 crores to Chennai Port to modernize and give uplift to Kasimedu Port.

"But nobody has come forward to take those Tender works", say the Port officials.

So the government should immediately tack back that huge amount and take up works through public works department. At present there is space to assemble 250-300 motor boats. But 1300 motor boats are being erected and no sufficient space is available. The boats rub each other in that congested space and no repair work is done in a smooth manner. There is often fight for space between the laborers engaged in boat erection works. Damages are observed due to dashing of boats with each other and it is expensive to set right it.

Poor fishermen find it difficult to survive in that limited space.

The minor Ports are considered as back bones of all trade developments and it is high time that a team of experts analyze deep into the problems and find a way or it.

****

Shipping &Maritime Expo 2014 Exhibitors

Shipping & Maritime Expo 2014 took place in NCPA complex from December 11th to 13th.

Totally there were 27 participants.

Delegates and visitors in thousands enjoyed the show on all the 3 days. Especially the international traders, who were interested to have tie-ups and collaborations, were seen busy in discussions for further business developments. The Marine Institute students had a good time to observe the latest trends and innovations in maritime sector.

On the whole it was a useful exposure to all participants.

Bharati Shipyard, the leading ship builder in India took a good opportunity to reveal their greatness in this field, through this informative INMARCO-INAvation 2014 Expo 2014, held at NCPA Complex in Narimanpoint, Mumbai.

This company listed in Bombay Stock Exchange and NSE has won laurels in taking up ship erection works in Mumbai, Kolkata, Goa, Ratnagiri, Mangalore and Dabhol.

In Mumbai, it is engaged in steel fabrication and major piping works on 12 acre land with CNC Plasma cutting, Panel line fabrication and Auto welding machines. It is located in Ghodbunder.

Bharati Ship Yard Ltd has the capacity to build hulls up to 125mt length in Kolkata, vessels up to 150mt length in Goa, vessels up to 220mt length in Mangalore and in Dabhol Yard. It can erect Jack up – Rigs, Off-shore structures and vessels up to 100,000 tons and 220mt length. Dabhol Yard is the largest among all with an area of 30 acres.

They have been asked to erect 150T DP – 2 AHTVSVs, 220t Diesel Electric DP-2 AHTSVs, 100m Diesel Electric DP-3 MSVs, 80t DP-2 AHTSVs, 4000T DWT DP-1 MPSVs and 3300TDWT DP-2 MPSVs.

With 3 level quarters for 110 individuals, Heliport, 3Nos of PCM-120ss cranes, 60 nos of 375 –Kip Electro mechanical Rack and Pinion Drives, Letourneau Super 116E class platform 65T,80T to 120T, Bollard Pull Fire Fighting, Anchor Handling Tug Cum Supply Vessels etc were all erected by this reputed ship building company. Diving Support Vessel, Offshore Crew Boat and 100 m Tanker cum RORO vessels were all fabricated and erected by this company.

Contact address: Bharati Ship Yard Ltd,

302, Wakefield House, Sprott Road,

Ballard Estate, Mumbai – 400 038. India.

Phone: +91-2230289200.Fax: +91-2230289222.

Saifee Enterprises, which imports aircraft tools, power tools, offshore tools, wood working tools, automobile tools and industrial cutting tools, had occupied a space in this expo. Contact person: Hunaid Lokdandwala. Email id: saifee.ent@gmail.com, Cell +91-9867645115.

Contact address: Saifee Enterprises,

44, Nagdevi, X lane, 2nd floor,

Office No.20, Mumbai, 400 003, India.

GOL Offshore has its office in Energy House, 81, Dr.D.N. Road, Mumbai-400   001.   Phone: 91-22-66352222, email:bdmoff@goloffshore.com. They exposed their various activities in this great exhibition. For the past 32 years they have been engaged in marine logistics, drilling, marine and offshore engineering ship repair maintenance, port and terminal support etc.

Not only in India, they have served in middle east with 4 offshore construction support infrastructure, 2 construction of Hook up Barges and construction support for Work Boats. They have served with PSV, AHTSV, OSV and AHT in JNPT, Tutucorin Port, Kolkata Port and private ports like Gujarat Adani Port, Gangavaram Port and Pipava Port etc.

"Rescue Operations done by them for Indian Naval Frigate, Product Tanker in Mumbai Port and General Cargo vessels erected by them are lauded by all", said their stall staff in this show.

In Mumbai it has its own 2,400mt floating Dry dock.

TEBMA Ship Yard Ltd, Malpe Harbor Complex, Malpe, Udupi, 576106, Phone:+91-8202538900 occupied a space in this show to exhibit their potentials in erecting Multipurpose Support Vessels, Cutter Suction Dredgers, Utility Vessels, Geo Technical Research Vessels, Harbor Tugs and Special Purpose Customized Vessels. Their corporate office is in 40, Bazullah Ro, 2nd floor, M.T.Rajen Properties, T Nagar, Chennai – 600 017, T Nadu. Phone: 91-44-28340703, Fax: 91-44-28340702.   Email: info@tebma.com.   Email: info@bharatishipyard.com.

An agency is always needed to represent the grievances of, Ship owners, operators in Logistics, Manufacturers of Dredgers and Offshore Support Vessels, companies involved in Cargo Ships, Port and Terminal Operators. So, ICC shipping Association does it to solve issues with the government departments, environmental problems and tax concessions. They had a stall to explain all the above activities. Contact person: Jyoti Birare. Address: ICC Shipping Association, Scindia House, Basement, N.M.Marg, Ballard Estate, Mumbai 400001,   Phone.022-22623003,   22623912.Fax: 022 22623911,Email:mail@iccsa.in.

Chugoku Paints (1) P Ltd with their office in 405, Raheja Chambers, Free Press Journal Marg, Nariman Points, Mumbai – 400 021, Phone:022 - 43550600, Fax: 022 - 43550625, email: sales@cmpindia.net. Environment friendly, low labor expenses, longevity and fuel saving are some of the special features of their Seaflo neo, said the representative in the stall allocated for this company. Ramesh Tripathi, their Regional manager can be contacted by his mail id: rtripathi@cmpindia.net. Mobile: 09819943312.

Taurus Marine Services (P) Ltd, 402, Vidarbh Vaibhav, 64, Bhavanishankar Road, Dadar (W), Mumbai-400028. Phone: 91-22-24391700, Fax: + 91-22-24391719, email:tms@taurus-marine.com showed the Induction Buffet Warmer, 100Lts Kadai, Bulk Cooking and Fulka Puffer. They say that 16kwh Induction Heater will save Rs499800 per year and recommended for ships to save fuel and cost instead of using LPG burner.

Taurus Marine Services take up replacement of worn out marine parts in ship.

F.O., D.O & LO purifiers, Centrifugal Pumps, Flow Meters, Sludge Pumps, all types of Compressors etc are handled by them.

Mrs. Anjali Bhide is the Managing Director of this esteemed company, which has won ISO 9001; 2000 DNV certificate.

It seeks international customers for support in all service activities needed for marine industry.

Aryatech Marine & Offshore Services P Ltd, B-1, Hauzkhas New Delhi – 110016, Mob: 91-9999009126, Phone: +91-11-46018102, Fax: +91-11-46018103, Email:info@aryatech.net, was there in the show to tell the public about SACS (Structural Analysis Computer System), a product of Bently Systems. USA, Auto plant, Multiform, Auto pipe, STAAD PRO etc are some of the software they deal.

Indian Navy utilizes Max Surf to erect floating structures, calculate stability etc.

They render training in Subsea Systems Mooring, O&M of Offshore structures etc.

Naval Architecture Drill Ships, Semi-submersibles and Single Point Mooring are some of the areas they serve.

Essar Trading in Mumbai supply Piston type valve, Gate valve, Check valve etc. Pressure ranging from class 150 to class 2500, various sizes from 15mm to 250mm are marketed by them.

They are manufacturers and suppliers of Eccentric Wheel Assembly 200m to 260mm with all spare parts regarding it. Rescue

Boat Embarkation Light, Inclinometer, Electrical Cables, Plastic Cable Tie, Air Horn, Emergency Light etc are some of the products they deal with. Their office is located in Essar Trading & Engg Co,44, Nagdevi Cross Lane, 2nd floor Room 20, Mumbai -400 003.

Contact Person: Ravi Mehta, Jivan More. Mobile Phone: 9967556022, Land Line: +91-022 23420484, 022 22136537.

Ken Mark Tech solutions with branches in UAE, New Delhi, Goa, Surat, Bangalore and Chennai, has its registered office in Mumbai.

They render services for Oil pumps, Air compressors, Emergency Engines, Heaters, Boiler and Turbine set for power generation from 0.5mw to 2000mw.

Repair works in cracked engine casings, metal stitching in engine cylinder head, reconditioning of any part of the ship etc are carried out by them.

Timely delivery competitive rates and talented technical team are plus points for this company, said their representative.

Their workshop is located in A-3/A-4, Singh (1) Industrial Premises Co-op Society, Ram Mandir Road, Goregaon (W), Mumbai, 400 104. India. Email: info@kenmark.in.

Aries Technical Service P Ltd is located in D-2, Udyog Sadan-3, Maroli M.I.D.C, Andheri (E), Mumbai, 400093. Phone: +91-2240271800, Fax: + 91-22-400271818, Email: info@ariestech.co.in. They deal with clutches manufactured by Quick Shift USA. They are interested in dealing with other marine equipments, sales and service with accuracy.

NUSI Offshore Training Institute had a stall to tell the visitors about the training they impart to serve to Maritime Industry efficiently with good DP Simulator, Crane Simulator, ROV Simulator, OSV and Engine Simulator. The students gain abundant knowledge practically to face any problems on ship. Excellent Cafeteria, Gymnasium and green garden make the students study with all relaxation and joy. Their office is located in NUSI Offshore Training Institute, Thakurwadi

Road, Wardoli Village, old Mumbai, Pune Highway off Shedung Phata, Panvel, 401206. Email: nofipanvel@gmail.com.

Class NK, which deals with safety, environment, education, training and prime management had a spacious in this expo to tell the visitors about their services and products.

Software dealing with operation, maintenance and machineries of ship are their valuable services.

It has a worldwide service network in China, Singapore, Dubai, London, New York and Rio de Janeiro. They are yet to open their office in Germany. Their service centres are in Algeria, Amba, Argentina, Australia, Bangladesh, Brazil, Canada, Dalian, Colombo etc and many more countries like Yemen, West Indies, New Orleans, Los Angles, Seattle, Ukraine, Uruguay, etc.

The exhibition conducted along the sea shore of Mumbai was informative and interesting.

More than 15000 visitors and delegates from all over the world made it a grand success.

****

Major ports in Andhra Pradesh

Vishakapattinam Port is a major port in Andhra Pradesh. It is located at lost 17°42' N and long 83°23'E It has a plan for Quay 1 and 2 for export facility for bulk cargoes. To handle fertilizer east Quay is to be remodified. Creation of east clock and expansion works in it to enhance east Quays. Deepening of entrance channel is another major plan in this important port. These above activities will be taken up in inner harbor area.

For outer harbour area improvements in iron are jetty and widening of container terminal are important plans. The general cargo berth is to be mechanized. The improved provision to handle crude oil and pol in this port is a great vision of the port management. Even though we have naval facilities in inner harbor, extension works have been suggested.

It has storage capacity of 2,50,000 tons for imported goods materials and goods are received from vessels which are 80,000 DWT. For export activities it receives goods via roads, railways and load in 50,000 DWT vessels.

As this port is an important major port amongst the 13 major ports in India, heavy handling and unloading takes place. Hence facilities are to be increased to store goods weighing 1,50,000 t.

It has in-motion weigh bridges to check the weights of the goods. This weigh bridge is a recognized and certified by govt of India and Indian Railways.

EQ-8 and EQ-9 are 255m long and they are Panamax compatible berths. Loading and unloading operations of the goods are done by 3 cranes which are capable of vertical discharge. These mobile cranes are to be modernized.

Rail weighs bridges with 100ton capacity-3 Nos. This is located within the terminal.

Road Weigh Bridge which is flat type-1 no. and weighing capacity of 80t

The vessels which have arrival draught -11m can be accommodated here.

To monitor malpractices and thefts cameras are fixed in vital locations in the port. Experienced security is posted all over the port terminal. As goods, like coal, lignite, cement, are handled entire area is sprayed to avoid dust free environment.

The port has good roads to transport goods in and out of the terminal.

The goods unloaded are stored in area with good and strong concrete with 200,000t capacity.

All details, reports, information about incoming and outgoing of vessels are passed to all divisions with a centralized information centre network.

Well trained manpower to monitor the port activities is available. Some are foreign trained. Manganese is caustic soda, Aluminum, cement, fly ash, food grains, pulses, calando petroleum iron ore. Lime stone, gypsum, pigriron, timber, coke, chemicals, electrical goods, fertilizers etc are handled here.

6 million ton of cargo handling is a great achievement of this port.

As they have earlier handled 6.27lakh mt in 30 days, now they have exceeded it and they aim to handle 6.58lakh/mt within a month. This is achievable with dedicated staff employed in this great port. It has discharged 7746 mt of stredded scrap within a day. This has attracted and increased the import activities in this port.

The natural deep water basins in outer harbor is a gift to this port to accommodated vessels up to 1, 50,000DWT and draft up to 17mts in the outer harbor, Inner harbor can accommodate 11mt draft and LOA up to 230mt

37150t of iron one was loaded recently at WQ-3. This is record says the port management major investment done in these ports are Rs.1080 core expenditure for container terminal and outer harbor works.

Works EQ-8 and EQ-9 with an estimate of Rs.1960 crores.

Deepening entrance channel and inner basin up to 14mt at a cost of Rs150crores is an important work to be taken up.

Strengthening of EQ ways to 14mt at a cost of Rs.120 crores and providing mechanized import facility at EQ-13 with an estimate of Rs.79 crores works in east dock is and northern side with an estimate of Rs.89 crores will give a face lift to this sophisticated port in Andhrapradesh.

Mechanized facilities for coal handling in East docks, installation of ship leader Alumina in WQ7, replacement of 2 locos 1430hs and 1 loco 3100Hs, up gradation of iron one jetty to 20,000 DWT are some of the special works taken up here. New Tugs 75TBP at a cost of Rs.90crores. SBM facility for handling of crude oil by VLCC at a cost of Rs.540 cores, strengthening of Quays to 125mt at a cost of Rs.47 cores are added works done in this port. Mangalore port is a major port in Karnataka. It is located at 12°55'N and 74° 48'E.

Being a busy port in Karnataka state, proposals for creating berth 15 in new western dock to handle coal, development of SBM facilities to handle crude oil imports, increasing the depth of channels, remodification of berth 1 and 2 for handling container goods, creation of container terminal in western dock, remodification of berth 13 to speed up the bulk quantity of liquid.

A proposal to expand the iron one storage capacity for berth 14. But it was abandoned.

Implementation of ERP, additional oil berth and second terminal erection activities in this port are significant improvements seen in recent years.

Procurement of sophisticated cranes, post based SEZ, deepening of lagoon, deep draft general cargo are on the anvil.

This port in southern side of India offers important sea routes to USA, trope, east Africa, East Asia and Middle East. It has achieved to handle more than 33 million tones/year.

Train links through Tamilnadu, Kerala, Mumbai,Coimbatore are vital for business inks. West coast Express, Trichy Express, Mangalore mail, Pondicherry express, Rajhani Express, ERS Pune Express, Mumbai Express, Happa Express, Massyaganga Express, Maru Sagar Express, Sampakranthi Express, Dehradun Express etc. So excellent rail links are available for Mangalore and the port located here. The Airport located here connects all cities 6 many countries.

The highway which links kochi and Mumbai covers many important business towns along it. This important highway is NH17 which leads to Mangalore port also. To have transportation of goods from Bangalore to this port, we have NH48. From Sholapur all important containers pass along NH13 to reach Mangalore port. 4 lane road works from Suratkal to Bantwal will be completed shortly and this will enhance more truck movements to this port.

Wooden log merchants are benefitted more from this port as 25015 tones are handled in a year. These are exported to Gulf countries and Europe. Cashnew is grown abundantly in this red soil of Mangalore town. Hence 1, 24,253 tones of cashew are handled and exported to other countries from this port.

12592T of chemicals, 14037T of salt and 10,097T of tiles are handled in a year.

Connectivity for Ports

Vishakapattinam port is connected to Chennai Howrah main railway off the East Coast route. Connectivity with NH5 is a big advantage for transportation of goods to this port. The project for gave a face lift to the 12.47km long Naval Dockyard and industrial by pass state highway is on the anvil.

Ennore port has good rail connectivity.

NH4, NH5 and NH45 are major highways for logistics operations.

Ennore-Manali road is added advantage for loading and unloading of goods at fast rate. Chennai Port Trust is linked with railway network to transport goods to all parts of Tamilnadu and other important cities like Mumbai, Hydrabad, Kolkatta, Gujarat, Ahmadabad and Karnataka.

NH45B, NH-7, NH-7A are vital roads to carry goods in and out of the Tutucorin port in T.Nadu. The four lane roads from Tutucorin and Tirunelveli known for a distance of 48km will add more traffic and port business via NH-7A. NH45B is another gift to Tutucorin port for excellent logistic movements.

Broad-gauge railway links to transport goods to Tirunelveli, Nagercoil, Trivandrum, Madurai, Bangalore, Chennai, and Trichy are plus points. Good rail link with Bangalore, Karur, Coimbatore, Tiruppur Salem which are busy industrial area. Tiruppur is popular for garment manufacturing u its and hence this port gives a big hand for its fortune.

Chennai Port is doing good business through NH5, NH4, and Nh45 to cover areas like Chennai-Kolkatta, Chennai-Bangalore/ Secundrabad, Chennai-Dindigul, and Karur respectively. On east coast we have excellent roads to transport goods to Pondicherry and southern districts of Tamilnadu.

Chennai Port-Maduravoyal 4 way lane elevated corridor plan adds more benefits to Chennai Port business. Cochin port, an important port in Kerala has a single railway line Shoran-Trivandrum supports the port goods movements. It branches off at Ernakulum from the main line.

Toad and Rail connectivity is available for international container Transshipment terminal to enhance import and export.

Two major bridges on Mattancherry channel and Ernakulum channel extends support for transportation of port goods by roads. A link road to NH47 by pass from willing Don Island is an import traffic movement path. 4 lane projects in NH47 is another big advantage to this major port.

New Mangalore port has good road links through NH48 (to link Bangalore and Mangalore), NH17 to cover Cochin, Goa and Mangalore, Nh13 to extend up to Sholapur from Mangalore.

Four lane of NH17 for easy traffic movement between Suratkal and Nantur, NH45 to extend transportation between Padil and Bantwal, by pass road on NH17 to extend good traffic movement between Nantur junction and Padil junction are special for logistic facilities in this port.

There are 2 railroad projects to be completed before 2017 in Vishakapattinam port. Another 3 more is planned to be executed in 2017-2020 for the same port. Total investment is Rs.750 crores. Ennore Port limited has 1 connectivity plan to be executed before 2017 with an investment of Rs.450 crores.

Chennai Port plans for 2 rail-road project to be completed within 2020 with an investment of Rs.225 crores. The commencement of this project will take place in 2017.

Tutucorin port has 1 major rail-road project with an investment of Rs.640 crores. As a phase-3 plan another connectivity works will be done in 2017-20 with an investment of Rs.300 crores. Cochin port has one rail road project to be completed with 2017 with an estimate of Rs.40crores.

Port of Cochin

Bunkering terminal, creation of cruise terminal, national highway and rail linking projects, erection of LNG and LPG terminals, development of SBM facilities for crude oil imported room various countries, improvement and facelift to willing don island are some of the plans on the anvil to improve business. While executing the road plans and new railway routes, border issues crop up with Tamilnadu. As two different parties rule Tamilnadu and Kerala, there is no understanding for this vital infrastructure.

Compared to other ports investment made from their own resources in Cochin port is 3% less. It is Rs.556 for a 3 year period while investment done in VPT is RS.4.379 crores.

Increases in revenue for Cochin for the period 2013-2014 is Rs.502 crores, whereas for TNPT it is Rs.1841 crores. For Mangalore it is Rs.345 crores. For Chennai port, Ennore port and Vishakapattinam it is Rs.735 crores, Rs.546 crores and Rs.755crores respectively. Project profit after tax for 2013-2014 for Cochin, Mangalore, Tutucorin Chennai, Ennore, Vishakapattinam is Rs.208 crores, Rs.134 crores Rs.55 crores, Rs.239 crores, Rs.403 crores, Rs.249 crores respectively. Investment undertaken in Tutucorin port for berth no7 5mtpa is Rs.1000 crores. For construction of NVW for NLC and Tamilnadu Electricity Board is Rs.490 crores with 6.30 MTPA.

For container terminal, outer harbor works Rs.1080 crores is pumped in. for erecting multipurpose berths EQ-8 and EQ-9 Rs.1960 crores is invested.

Rs.4690 crores for container terminal, Rs.4950 crores for development of second container terminal is inverted for Chennai port.

Rs.2490 crores for marine liquid terminal, Rs.3990 crores for coal terminal and Rs.4800 crores for iron ore terminal are major investments done in Ennore port.

For crude oil handling facility, international container transshipment terminal Rs 7200 crores, Rs21180 crores have been invested. For LNG-Requalification terminal Rs31950 crore has been invested.

As there is lack in inter- port competition, growth in facilities, road facilities, berth amenities and modernization is poor in all ports located in south India. Within port competition is also poor, when compared to TNPT which has 3 container terminals.

For handling of containers Ennore leads other ports in south, next is Chennai port.

Market share per port for coal is high in Ennore, while Cochin and Chennai and Mangalore are low. Vishakapattinam stands second and Tutucorin stand next to it.

Traffic for iron-ore coal, fertilizers and crude for new Mangalore port is 10mt, 20mt, 430mt, 209mt, 125 2720mt.For port of Cochin total throughputs is 2463mt for 2012-13. For Tutucorin port it is 30, 80mt and for port of Chennai it is 6417mt. Amongst the south Indian ports, Ennore port's total throughput is 40,64. It is observed that it will extend to 13640 for the period 2025-28.

Traffic projections for Vishakapattinam are observed to be 8170mt for 2012-13 and 14680mt for the period 2025-26.

****

Development of Minor Ports in Tamilnadu!

With private participation it has been decided to develop the minor Ports in Tamilnadu to enhance industrial and economical growth in Tamil Nadu.

This step is to be effectively implemented with liberal funds and efforts.

In Tamilnadu we have 3 major Ports like Chennai, Tuticorin and Ennore Port.

Apart from these Ports we have 15 minor Ports like Cuddalore, Nagapattinam, Rameswaram, Pamban, Kulachal, Valinokkam, Kanyakumari, Ennore, Punnaikayal, Thirukadaiyur, Py3, Kattupalli, Thiruchopuram, Manapadu and Kudankulam.

Considering the gains obtained from development of these Ports, the Ministry for Seaport is taking active steps with huge funds. By developing these minor Ports, we can expect growth in Thermal Power Stations, industries based on production of anchors, chains and sea foods. Oil Industries will also thrive with their support says the experts. In 2006 there were 15 minor Ports in Tamilnadu and it has increased to 20.

Cuddalore Port is at a distance of 200km from Chennai. Tamil Nadu Government planned to erect walls to avoid erosion, new administrative buildings and new wharfs on 100 acre area. Due to storage of funds the works were held up. So it was decided to handover the projects to private parties for development and maintenance for a 30 period of lease.

Chemplast, a group India Cements has created a Marine Terminal facility in Cuddalore to handle their Polyvenyl Chloride Chemicals.

Cuddalore Powergen Corporation has planned to erect storage facilities in Cuddalore port at an estimate of Rs.325 crores.

Chennai Petroleum Corporation has taken up Oil refinery plant erection and oil handling facilities in Nagapattinam Port.

Such participation of private agencies only can enable a sea-change in Port's infrastructure in Tamilnadu.

Present condition of Minor Ports

Kattupalli: -

This is near Ennore.

At an estimate of Rs 3375 crores ship building facilities and private Port erection are in full swing.

Mugaiyur: -

This is near Mamallapuram.

A small Port erection was announced by the Tamilnadu Government. A ship building facilities with Rs.500 crore estimate will be witnessed shortly.

Thiruchopuram: -

This is located in Cuddalore district.

Nagarjuna group has been permitted to erect a Port with an estimate of Rs.800 crore. Crude Oil and Petroleum will be imported and refinery works will be taken up by them.

Silambimangalam: -

Good Earth Ship Building a Private Company has been sanctioned to erect a Port here with an estimate of Rs.500 crores.

This is near Cuddalore District.

Ships weighing 75000T can be anchored in this Port.

Kaveri: -

Bell Power, a multicrore private company has planned to erect a 320 mw power plant in Poompuhar in Nagapattinam District.

To handle the required coal for this power plan, ultra modern facilities will be erected there by it.

This will give a big uplift for Kaveri Port.

Vanagiri:

This is located in Seerkazhi in Nagapattinam district, where a 500 mw Power Plant with an estimate of Rs.250 crore will be erected by NSL Power.

A big facelift to that Port will be given to handle the coal required for it.

Thirukadaiyur: -

This is also located in Nagapattinam district.

In Pillai Perumalnallur, PPN Power a private power generating company has started a 330mw Power generating plant.

To handle Naptha and natural gas for this plant Thirukadaiyur Port was erected in 1996.

Thirukuvalai: -

It is located in Vettaikaran Iruppu in Nagapattinam.

Tridem Port and Power Company has erected 2000mw power plant at a cost of Rs.650 crores.

To handle coal for this plant a Port has been permitted to erect on 276 acre land on lease.

Manapadu: -

This is located in Tuticorin and it was erected at a cost of Rs.800 crore.

Indian Power Project has planned to erect a 200mw gas based power plant near it and 100 acres of land has also been allocated for this huge project.

The Port will be supporting the needs of that power plant.

Kudankulam: -

Kudankulam Atomic Power Plant is located in Kudankulam (Tinneveli District).

For this power plant, arrangements to store sea water for a length of 1km and width 500mt are being executed.

With an estimate of Rs.340 crores Jetties and other amenities are being erected.

Panayur: -

In this Port a facelift is being provided to handle lignite and coal for 4000mw power plant to be erected by Coastal Tamilnadu Power Limited.

Parangipettai: -

This is developed by I.L. and FS Ltd to cater the needs of 4000mw power plant which is to be erected at an estimate of Rs.300 crore. Coal handling facilities will be provided here with all sophisticated equipments.

Udangudi:-

Chennai Udangudi Power Corporation Ltd has planned to provide facilities like wharfs, cranes etc for unloading goods from ships.

Also amenities to support 1600mw Super Thermal Station which is to be erected here will be provided.

Kasimedu Fishing Port

"Due to lack of maintenance, the Chennai Kasimedu fish Port will collapse within a year or two" say the specialists.

It is high time the authorities take some smart steps to protect it.

For all fishermen in Chennai, Tiruvallur and Kanchi District, this Kasimedu Fish Port has been an important Port to fish for their livelihood.

There more than 1300 motorized boats and hundreds of fiber boats are engaged in fishing. Around 30,000 men are engaged in this profession. Out of this 13000 are involved directly in fishing.

Indian earns a foreign exchange of Rs.10500 crores through export of fish and out of which Rs.500 crores is earned from Tamilnadu.

The fish export business has played a leading role in Tamilnadu's income.

When such is the situation it is surprising that State Government and Central Government ignores it with least care.

At times of election they speak high about budget allocation to develop fishing and fish export activities.

Several crores are also available but nothing is done usefully or fruitfully. So the age old fish Ports are still in uncared condition which is totally unhealthy for economic growth and also for environment.

Big holes and pits are seen in wharfs and 3 fishermen have slipped and fallen into it.

Yet no repair work is taken up to set right it.

Some temporary supports are seen which are not sufficient for safety and security.

Many areas where fish are auctioned are in unhygienic and filthy condition. Maintenance and upkeep are pitiably ignored. Even rickshaws cannot enter to carry diesel for the motor boats.

Basic amenities like lighting, toilet, etc are neglected. Even a decent restaurant could not be seen for fishermen to have a healthy food.

When crores are allocated for betterment works it is unbearable to see that area filled with pungent smell and rubbish.

Huge money is handled here and no government or private Bank is seen there to handle it and safeguard it. After each fish trading transaction is over it is unsafe to carry that lump sum in gunny bags all the way to a Bank in a distant location.

Due to improper lighting the diesel thefts are common here.

Sometimes boats are also stolen by miscreants.

The Tamilnadu government has handed over Rs.20 crores to Chennai Port to modernize and give uplift to Kasimedu Port.

"But nobody has come forward to take those Tender works", say the Port officials.

So the government should immediately tack back that huge amount and take up works through public works department. At present there is space to assemble 250-300 motor boats. But 1300 motor boats are being erected and no sufficient space is available. The boats rub each other in that congested space and no repair work is done in a smooth manner. There is often fight for space between the laborers engaged in boat erection works. Damages are observed due to dashing of boats with each other and it is expensive to set right it.

Poor fishermen find it difficult to survive in that limited space.

The minor Ports are considered as back bones of all trade developments and it is high time that a team of experts analyze deep into the problems and find a way or it.

****

www.ingramcontent.com/pod-product-compliance
Lightning Source LLC
Chambersburg PA
CBHW021847170526
45157CB00007B/2975